Highlights from Millions Fed

PROVEN SUCCESSES IN AGRICULTURAL DEVELOPMENT

David J. Spielman and Rajul Pandya-Lorch

IFPRI®
INTERNATIONAL FOOD POLICY RESEARCH INSTITUTE
sustainable solutions for ending hunger and poverty
Supported by the CGIAR

2020 VISION

Acknowledgments

The highlights presented here are drawn from the book entitled *Millions Fed: Proven Successes in Agricultural Development*, edited by David J. Spielman and Rajul Pandya-Lorch. We gratefully acknowledge the support of The Bill & Melinda Gates Foundation for this project. We warmly thank the members of the Millions Fed Advisory Committee and the authors of the Millions Fed case studies for their valuable contributions and insights. The initiative by all who submitted or nominated examples of success is greatly valued—there may be many more successes in the making.

We extend our appreciation to Joachim von Braun for his guidance and support. We are indebted to our colleagues on the Millions Fed Project Team—Klaus von Grebmer, Kenda Cunningham, and Sivan Yosef—and to Heidi Fritschel, Gwendolyn Stansbury, and Evelyn Banda for their tremendous dedication and enthusiastic involvement throughout this project. Thanks also go to Derek Byerlee and Steve Haggblade for their useful advice. IFPRI's Publications Review Committee and several anonymous reviewers provided thoughtful comments.

Updated November 2009.

DOI: http://dx.doi.org/10.2499/0896296598

ISBN: 10-digit: 0-89629-659-8
13-digit: 978-0-89629-659-6

Photo credits

Cover images from left to right: © Richard Lord, © iStockphoto, © CRS/David Snyder, © Jan Banning/PANOS, © CRS/David Snyder, © Klaus von Grebmer/IFPRI.

Cover design/illustration and booklet design/layout: Joan K. Stephens/JKS Design.

Page iii: © One Acre Fund. Page 2: © Jeremy Horner/PANOS. Page 3: © Yue Jin/USDA. Page 4: © Giacomo Pirozzi/PANOS. Page 5: © Intercooperation. Page 6: © Alfredo Caliz/PANOS. Page 7: © Workneh Tibebu/IFPRI. Page 8: © Tom Pilston/PANOS. Page 9: © AVRDC-The World Vegetable Center. Page 12 top: © National Dairy Development Board. Page 12 bottom: © Kevin Fitzsimmons/Univ. of Arizona. Page 13: © Chris Stowers/PANOS. Page 14: © Helen Keller International. Page 15: © Helen Keller International. Page 16: © Latha Nagarajan/IFPRI. Page 17: © Helen Keller International.

Learning from Success

The world needs to greatly accelerate its progress in reducing poverty and hunger. At present, one in six people worldwide suffers from hunger and malnutrition—a tragically high proportion—and many more cannot afford a healthy diet. And as progress is being made, more challenges are on the way: the world's population is projected to grow to 9 billion by 2050, climate change is raising risks for farmers, environmental degradation is contributing to poor soils and scarce water, and we still face the same problems that led to devastating volatility in food prices in 2008.

It is also important to remember that the world has already achieved great successes in agricultural development that have fed billions. After all, although a grim Malthusian world once seemed inevitable, some 5 billion people now have enough food to lead a healthy and productive life and the proportion of people who are hungry is falling. The experiences of success that led to this achievement may offer valuable lessons about how to put agriculture to work to solve hunger and malnutrition. Until now, however, relatively little evidence has been available on where, why, and how these interventions succeeded.

To identify and examine successes in agricultural development and draw out the lessons they offer, The Bill & Melinda Gates Foundation (BMGF) called upon the International Food Policy Research Institute (IFPRI) to assess the evidence on what works in agriculture—what sorts of policies, programs, and investments in agricultural development have actually reduced hunger and poverty. This project follows on another recent project supported by BMGF and led by the Center for Global Development called "Millions Saved: Proven Successes in Global Health."

The case studies of success were chosen through a rigorous process that included an open call for nominations, a wide-ranging literature review, and expert consultations. More than 250 candidate case studies were winnowed down using a comprehensive set of criteria that took into account such issues as scale, impact, and sustainability. A committee of recognized international experts provided valuable insights and advice. Ultimately, the project identified 20 proven successes. These spanned from interventions enhancing productivity to combating diseases and pests, conserving natural resources, expanding market opportunities, improving human nutrition, and improving the policy environment. A common thread running through many of these success stories is the confluence of science, policy, and leadership.

Until hunger and malnutrition are eradicated, success cannot be truly claimed. Our hope is that this effort will direct more attention to sound agricultural development investments that cut hunger and to facilitate the scaling up and replication of successes.

Joachim von Braun
Director General, IFPRI

Prabhu Pingali
Deputy Director, BMGF

Farmer applies fertilizer to maize, Kenya.

Fifty Years of Progress

In the late 1950s around a billion people—about one-third of the world's population—were estimated to go hungry every day. Famines were threatening millions in Asia and Africa in particular, and prospects for feeding the world's booming population looked bleak. In response to this alarming picture, scientists, policymakers, farmers, and concerned individuals initiated a concerted push to boost agricultural production and productivity in developing countries. Developing and industrialized countries, together with development agencies and civil society organizations, pursued a range of interventions in agriculture: they applied modern science to crop and livestock production, constructed irrigation systems, developed new cultivation practices to conserve natural resources, introduced policies to encourage farmers to grow and sell more food, and launched many other programs in agricultural development. The result? About one billion people now go hungry every day.

This result may look like failure, and in one sense it is. The fact that 1 billion people remain hungry and malnourished is a tragedy on a grand scale. Looked at another way, however, the present situation reflects astounding success. While the absolute number of people who are hungry has remained the same, the relative figure—the proportion of the world's population that has remained hungry—has declined dramatically. In the mid-1960s, when the global population was about 3.3 billion, only about 2 billion people were getting enough to eat. Today's population has burgeoned to more than 6 billion—and some 5 billion people now have enough food to live a healthy and productive life.

Efforts to increase the global availability of food have led to enormous gains in agricultural productivity and food production, with yields of many staple crops multiplying severalfold. Great strides also have been made in improving the quality of food so that it contributes to good nutrition, and in improving the ability of the most vulnerable groups—women and children most significantly—to access food needed for survival.

Importantly, these efforts have done more than just feed millions. The interventions of the past half century have also demonstrated that agriculture can be a key driver of growth and development for many of the world's poorest countries.

While the causes of chronic hunger and persistent malnutrition are complex, the experiences of the past 50 years show that the solutions are by no means beyond our reach. But what do we really know about what works in agricultural development, and where, when, and why some interventions succeed? Which policies, programs, and investments in agricultural development can substantially reduce hunger and malnutrition? And which of these interventions can do so within a changing global landscape characterized by growing natural resource scarcities, climate change, global market volatility, and major health and demographic changes?

Family harvests wheat, India.

What worked?

Successes in developing-country agriculture are extremely rich and diverse in nature, varying in time, space, and character. Some are successes that emerged for just a few short years to trigger long episodes of growth and development. Some have emerged from years of dogged persistence that yielded returns despite substantial risks, uncertainties, and doubts. Others have resulted from communities that took action to ensure their own survival under difficult environmental conditions. Still others are successes that were inspired by leaders and organizations who marshaled the resources needed to contain the spread of crop and livestock diseases that know no boundaries.

The pathways to success are also extremely varied. Some cases demonstrate how an improved crop variety or cultivation practice contributed to improving food security by increasing crop output per hectare of land, lowering production costs, or reducing crop losses caused by pests, diseases, drought, or soil erosion. Others demonstrate how new agricultural technologies improved the sustainable use of scarce resources like fertile soil and water, or enhanced the nutritional quality of food that people both cultivate and consume. Still others illustrate how changes in incentives—whether public policies, commercial regulations, or socioeconomic norms—encouraged farmers to produce more food, pursue more sustainable cultivation practices, and participate more actively in the marketplace.

But these pathways to success are not simply about increasing the physical supply of food. Rather, they are about reductions in hunger that result not only from an improvement in the physical availability of food, but also from a change in an individual's ability to secure quality food. This change may result from any number of situations: an improvement in an individual's ability to produce food within the farm household; an increase in income that provides a consumer with greater purchasing power in the market; or a shift in norms that reduces the impact of practices and behaviors that limit an individual's entitlement to food within the household, community, or society.

We look at successes in six different areas:

1. intensifying staple food production;
2. integrating people and the environment;
3. expanding the role of markets;
4. diversifying out of major cereals;
5. reforming economy-wide policies; and
6. improving food quality and human nutrition.

1 *Intensifying staple food production*

A loose timeline of recent successes in agricultural development begins somewhere in the mid-twentieth century, when the menace of war, hunger, and disease loomed large for many developing countries that had just gained independence from colonial control or influence. A key driver in these early successes in agriculture was crisis—whether the result of human actions such as conflict, oppression, or complacency, or the result of natural causes such as drought or pests.

During the late 1940s and early 1950s, astute political leaders keenly recognized that hunger was a threat to long-term security, development, and prosperity. The real and perceived threat of famine ushered in an era in which policymakers' key priority was to increase the output (greater production) and yields (greater production from a given area of land) of staple foods.

One of the first major successes came from a global effort to fight wheat rusts—a plague that has been known to humanity for thousands of years but had never been effectively contained. Wheat rusts are actually fungi that can rapidly decimate wheat as it matures in the field, and are thus a threat to food security in industrialized and developing countries alike. The late Nobel Prize Laureate Norman Borlaug, with the eventual backing of policymakers, scientists, and philanthropists, catalyzed a global effort to combat the scourge by bringing modern science to bear on the problem—by breeding rust-resistant wheat varieties in Mexico with the help of innovative research methods. This global effort helped protect about 117 million hectares of land under wheat cultivation from wheat rusts, directly ensuring the food security of 60 to 120 million rural households and many more millions of consumers. Importantly, this global effort also secured a place for science and technology in developing-country agriculture and gave rise to a global agricultural research system, including the Consultative Group on International Agricultural Research, dedicated

Wheat infected with stem rust.

Processing cassava, West Africa.

to finding scientific solutions to ending hunger and food insecurity.

The wheat rust success evolved into a much larger and more multidimensional series of successes that began in the 1960s and came to be known as the Green Revolution. In Asia, this revolution began with the introduction of improved rice and wheat varieties for irrigated land that could be cultivated twice a year instead of once. The process continued into the 1990s as successes expanded to lesser-known staple crops such as millet and sorghum and more marginal areas dependent on rain rather than irrigation. The investments in science and technology—along with complementary investments in irrigation systems, road networks, fertilizer production, and food price stabilization policies—that underwrote the Green Revolution paid off handsomely. Farmers rapidly adopted the new farming practices and technologies to such a massive extent that between 1965 and 1990, cereal output and yields doubled, pulling India and other Asian countries back from the brink of famine. Between 1970 and 1990, an estimated 1 billion people benefited from the Green Revolution in terms of improved access to food, increased earnings from agriculture, or both.

Successes in Sub-Saharan Africa were smaller in magnitude but no less important in addressing the persistent threat of hunger in the region. In East and Southern Africa, applications of modern science to improve maize led to growth in both maize output and yields among the region's primarily small-scale, resource-poor farmers. Between 1965 and 1990, maize yields in Kenya, Malawi, Zambia, and Zimbabwe increased annually between 1 and 5 percent—rates that compare respectably with yield and production growth rates in countries such as the United States—while annual maize production increases ranged from 1.8 to 3.3 percent in these same countries.

In West Africa between 1971 and 1989, the application of modern science similarly helped contain the spread of a cassava mosaic virus (a disease) and mealybug (an insect). Both threats can generate major losses for cassava, a crop that is central to the sustenance and incomes of the region's poorest farmers, particularly in times of drought or crisis. By breeding cassava varieties that were resistant to the mosaic disease and by introducing a parasitic wasp to destroy mealybug in countries such as Nigeria and Ghana, the potential damage posed by these two threats was effectively contained. The introduction of disease-resistant cassava varieties is estimated to have contributed to making an additional 1.4 million tons of cassava flour or *gari* available per year, enough to feed 29 million people in the region. Similarly, the mealybug control program is estimated to have reduced losses from infestations by an estimated 2.5 tons per hectare.

2 *Integrating people and the environment*

By the 1970s, concerns emerged over the equity and environmental implications of rapid agricultural development. These new concerns encouraged a move away from a strictly yield-increasing outlook on food staple productivity to a more complex perspective on agriculture and rural development. Sustainable development issues came to the forefront of the development discourse, partly in response to issues accumulated during the Green Revolution, such as the overuse of agricultural chemicals, the depletion of scarce water resources, and the neglect of farmers' input into policymaking. New policies, programs, and investments were specifically designed to integrate rural communities into decisionmaking processes

Participation in community forestry, Nepal.

about their own development as a way of addressing sustainability along with equity issues. The idea that agricultural development could work if driven by direct community participation, environmentally sustainable cultivation practices, and supportive public policies gained a global following.

Experiences in Nepal that began in the 1970s illustrate this change in perspective. During this period, a series of prescient legislative reforms and innovative forestry programs contributed to a transformation of the country's strictly conservation-focused approach to its natural forests into a more broad-based strategy that encompassed forest use, enterprise development, and livelihoods improvement with direct benefits for the rural poor. Partly as a result of these reforms and programs, an estimated one-third of Nepal's population is participating in community forestry activities and directly managing over one-fourth of Nepal's forest area as a means of improving household food security and livelihoods.

In Burkina Faso and Niger during the 1980s, the rediscovery of community-based knowledge in the form of traditional agricultural management practices helped to transform the Sahelian region's arid landscape into productive agricultural land. In the wake of repeated droughts, and with technical support from charismatic community leaders and nongovernmental organizations, farmers began innovating on simple practices: protecting and managing indigenous trees and shrubs among crops to provide fodder and firewood and to improve soil fertility; digging pits on barren, degraded land to concentrate organic manure and rainwater for planting; and constructing stone contour bunds to control rainfall and runoff and combat erosion. In Burkina Faso's Central Plateau, the rehabilitation of between 200,000 and 300,000 hectares of land translated into roughly 80,000 tons of additional food per year, or enough to sustain about a half-million people in the region. In southern Niger, similar efforts are estimated to have transformed approximately 5 million hectares of land, improving food security for at least 2.5 million people.

Mother and child farming together, Kenya.

In Argentina, large-scale farmers adopted a different set of resource-conserving cultivation techniques, resulting in a significant increase in the global production of soybean in particular. During the 1980s, farmers, researchers, extension workers, and private companies worked together to promote zero-tillage cultivation—a crop management technique in which farmers essentially plant seeds in unplowed fields to maximize the gains from intensive double cropping and to lower production costs, with the added benefits of reducing land degradation, conserving soil fertility, and economizing on scarce water resources. By 2008, the area of land under zero tillage reached nearly 22 million hectares. The use of zero tillage, along with the introduction of herbicide-resistant soybean varieties and other factors, improved soil fertility by reversing decades of erosion, created an estimated 200,000 new agricultural jobs, and provided the international market with new supplies of soybeans that contributed to keeping global food prices low.

During roughly the same period of the 1980s, small-scale farmers in the Indo-Gangetic Plains—a vast region that encompasses parts of India, Nepal, Pakistan, and Bangladesh—began experimenting with similar zero-tillage cultivation techniques. An estimated 620,000 wheat farmers have adopted some form of zero-tillage cultivation since these experiments began, accounting for about 1.8 million hectares

Trading healthy cattle, Ethiopia.

of land in the Indo-Gangetic Plains and generating average income gains of US$180-340 per household, particularly in the Indian states of Haryana and Punjab.

3 *Expanding the role of markets*

In spite of these successes, many developing countries still suffered from slow growth, general economic malaise, and persistent food insecurity throughout the 1980s. A shift to more market-driven development took hold in many countries during this period. In some countries, this shift came in the form of structural adjustment programs that sought to rein in public deficits, improve national balances of payments, liberalize markets, and encourage private investment in the economy. In other countries, this shift occurred after the recognition that efficient supply chains played an important role in improving the production incentives for farmers, increasing incomes from farming, and improving food security. Market forces were expected to contribute to agricultural development, for example, by freeing up seed and fertilizer markets from state-owned monopolies, by removing price-setting policies in agricultural commodity markets to encourage more vibrant trading, and by closing the supply chain gaps that link farmers to markets.

In Bangladesh, government moves to liberalize agricultural input markets in the 1980s led to an easing of restrictions on the importation and sale of irrigation equipment, such as low-lift power pumps and shallow tubewells. These seemingly minor reforms stimulated the rapid growth of irrigated dry-season rice farming, which subsequently grew

Freshly milled rice, India.

to account for 90 percent of the increase in rice production in Bangladesh between 1988 and 2007. And with this growth in rice production came a decline in the real rice prices facing food-insecure households and ultimately significant reductions in poverty in the country.

In China, policy reforms that promoted private investment in agriculture, along with breakthroughs in rice research, fostered the growth of a vibrant seed industry for hybrid rice. Hybrid rice in China spread so quickly that between 1978 and 2008, it had grown to account for 63 percent of all land under rice cultivation. Importantly, its yield advantages helped China to feed an additional 60 million people per year during this period.

In India, similar policy reforms and scientific advances in the mid-1990s encouraged the growth of private investment in the marketing of improved seeds for pearl millet and sorghum, including hybrids. These two crops are essential sources of sustenance and income for some 14 million poor households in India. Although together they account for just 10 percent of the total cropped area in India, they are cultivated in the country's arid and semiarid regions where nearly 60 percent of the rural population lives. The emergence of private seed companies, combined with good public research, has provided an estimated 6–9 million farmers with access to improved seeds that have increased yields by up to 85 percent in recent decades.

Reforms in Burkina Faso's cotton sector that began in 1992 brought together experiences from both market liberalization and cash crop development as drivers of success in agricultural development. Saddled with a state-led cotton development strategy that was branded as inefficient, inequitable, and destabilizing to the national economy by the late 1980s, Burkina Faso pursued a reform path that combined efforts to strengthen the role of cotton farmers' groups before partially liberalizing input and output markets. Partly as a result of these reforms, and despite consistently low world prices for cotton, Burkina Faso has emerged as the leading African exporter of cotton based on a threefold increase in production since the early 1990s. The cotton sector's growth has absorbed more than 200,000 new farmers who were either engaged in the cultivation of other crops or were return migrants from neighboring countries experiencing civil strife.

In Kenya, policy reforms in the early-1990s contributed to the rapid growth of private investment in fertilizer and maize marketing, the outcome of which has been a dramatic reduction in time, effort, and costs associated with purchasing fertilizer and selling surplus maize production. The average distance that small farmers had to travel to purchase fertilizer decreased by half between 1997 and 2007, with similar decreases observed in the distances traveled to sell maize. The proportion of small-scale farmers using fertilizer on maize during the main growing season rose from 56 percent in 1996 to 70 percent in 2007, contributing to an increase in both yields and availability of this vitally important staple crop for Kenyan consumers.

4 *Diversifying out of major cereals*

The emphasis on markets also opened up new opportunities for cultivating and marketing non-staple crops—commodities such as legumes, fruits, and vegetables as well as dairy, livestock, and fish—as a means of increasing farm incomes and improving food security among the poor. Each success offers a different angle on how small-scale farmers, entrepreneurs, and policymakers responded to growth in market opportunities.

Across a range of Asian countries, the move away from food staples was exemplified by the diffusion of improved mungbean, a little-known pulse crop that is high in protein, iron, and other micronutrients, and particularly useful in maintaining soil fertility. Thanks to an international research program and active farmer participation in the research process, a wide range of mungbean varieties was released beginning in the mid-1980s, with traits such as higher yields, shorter maturity times, and other qualities that targeted a variety of agroecological conditions in the region. These improvements contributed to yield gains of 28 to 55 percent among an estimated 1.5 million farmers and were key factors in the 35 percent increase in Asian mungbean production between 1985 and 2000.

Global efforts to control and eradicate rinderpest—a livestock disease that, in its severest form, is capable of killing 95 percent or more of the animals it infects—reiterate the importance of livestock to rural livelihoods and food security. Concerted global, regional and national

Mungbean cultivation, Asia.

MillionsFed

PROVEN SUCCESSES IN AGRICULTURAL DEVELOPMENT

www.ifpri.org/millionsfed

South America

- **Innovating in the Pampas**
  Zero-tillage soybean cultivation in Argentina

Africa

- **Resisting Viruses and Bugs**
  Cassava in Sub-Saharan Africa

- **Re-Greening the Sahel**
  Farmer-led innovation in Burkina Faso and Niger

- **Navigating through Reforms**
  Cotton reforms in Burkina Faso

- **Unlocking the Market**
  Fertilizer and maize in Kenya

- **Breeding an "Amaizing" Crop**
  Improved maize in Kenya, Malawi, Zambia, and Zimbabwe

Global

- **Fighting a "Shifty Enemy"**
  The international collaboration to contain wheat rusts
- **Conquering the Cattle Plague**
  The global effort to eradicate rinderpest

Asia

- **Transforming Agriculture**
  The Green Revolution in Asia
- **Counting on Beans**
  Mungbean improvement in Asia

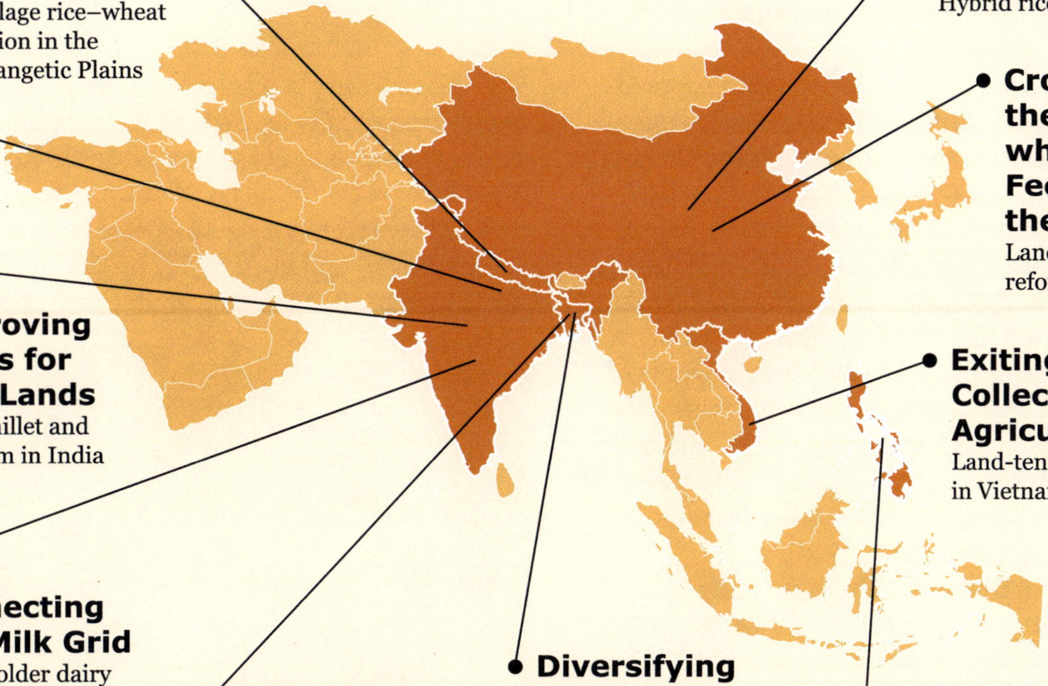

- **Leaving the Plow Behind**
  Zero-tillage rice–wheat cultivation in the Indo-Gangetic Plains
- **Seeing the Forest Through the Trees**
  Community forestry in Nepal
- **Pushing the Yield Frontier**
  Hybrid rice in China
- **Crossing the River while Feeling the Rocks**
  Land-tenure reform in China
- **Improving Crops for Arid Lands**
  Pearl millet and sorghum in India
- **Exiting from Collective Agriculture**
  Land-tenure reform in Vietnam
- **Connecting the Milk Grid**
  Smallholder dairy in India
- **Pumping up Production**
  Shallow tubewells and rice in Bangladesh
- **Diversifying into Healthy Diets**
  Homestead food production in Bangladesh
- **Farming the Aquatic Chicken**
  Improved tilapia in the Philippines

Smallholder dairy producers measure milk, India.

efforts in recent decades to control the spread of rinderpest through cattle vaccination, quarantine measures, and disease surveillance have played an important role in securing the livelihoods of small-scale farmers who keep livestock, as well as pastoralists whose livelihoods depend primarily on the health of their herds. Programs operating in Asia and Africa have helped to avoid potentially massive financial losses in terms of milk, meat, animal traction and, for many pastoralists, their main livelihood assets, and have brought rinderpest to the edge of eradication, the first time an infectious disease has been eradicated since smallpox in humans.

In India, Operation Flood, an innovative national program that ran from 1970 to 1996, helped create a national dairy industry that integrated small-scale farmers—many of them women—with village-level dairy cooperatives, commercial dairy processors and distributors, and new technologies to modernize the industry. With the backing of a supportive policy environment that ensured the dairy industry's steady growth and development, India went from being a net importer of dairy products to a major player in the global dairy market. Between 1970 and 2001, dairy production in India increased at the respectable rate of about

Harvesting farmed tilapia, Philippines.

4.5 percent per year, with estimates during 2007–08 indicating that dairy production has exceeded 100 million tons per year. As a result, millions of consumers now have better access to milk and other dairy products.

In the Philippines, the Genetic Improvement of Farmed Tilapia (GIFT) project that ran from 1988 to 1997 played an important part in enhancing the role of fish as a source of income and protein for many farmers and consumers. By breeding a tilapia strain that originated in Africa, the project developed a new strain that is faster growing and more resistant to environmental stresses than other strains. These improvements significantly boosted fish yields and output, thus increasing the availability of fish for consumers, reducing market prices, and providing a cheaper source of protein for the country's poor.

5 *Reforming economy-wide policies*

As the emphasis of agricultural development evolved and diversified over recent decades, the role of the agricultural sector in the wider economy has similarly changed. Economic policy reforms in recent decades have contributed significantly to changing the traditional urban biases that historically discriminated against farmers and, ultimately, against the poor. In some cases, trade and fiscal policy reforms have changed how both trade and aid are leveraged for development, transforming dependencies on food aid into more effective, long-term opportunities for development financing. In other cases, monetary policy reforms have reduced the distortionary effects of exchange rates and lending policies on the agricultural sector, allowing for more rapid growth and development.

The most dramatic case in point comes from China. Between 1978 and 1984, China undertook a series of policy reforms that transformed the country's food and agricultural sector and reduced hunger on a scale unrivaled in history. The reforms effectively reintroduced household farming after more than 30 years of collective agriculture. This new approach to agriculture—the Household Responsibility System—gave farmers the incentive to sell their surplus farm production to the market. By returning more than 95 percent of China's farmland to some 160 million farm households, the reforms directly contributed to an increase in rural incomes by 137 percent, a reduction in rural poverty by 22 percent, and an increase in grain production by 34 percent. Gains in on-farm efficiency also led to a substantial increase in the rural labor force available for non-agricultural employment, a shift that fueled a rapid process of industrial growth in rural China, and more broadly, China's remarkable march to industrialization during the past three decades.

In Vietnam, a series of similar reforms between 1987 and 1993 fundamentally shifted the country's economy to a greater market orientation, immediately transforming the agricultural sector. During the period 1989–92, the agricultural sector emerged from its stagnation and grew at a rate of 3.8 percent per year, while the country shifted from being a net food-importing country to the world's third largest exporter of rice in 1989. Within

Farmer harvests rice, Vietnam.

a decade, more than 10 million households—representing about 87 percent of peasant households—had received land-use certificates for about 78 percent of Vietnam's agricultural land. These reforms, together with other market liberalization policies, encouraged farmers to produce food staples, livestock, and high-value crops far more productively, and for substantially greater market gain, than in previous eras. The reforms contributed substantially to Vietnam's dramatic reductions in poverty and contributed to both economic growth and industrialization.

6 *Improving food quality and human nutrition*

While massive gains in improving the availability of and access to food were achieved in China, India, and many other developing countries as a result of these successes, far less has been achieved in improving the quality of food. Scholars have argued that the decades-old effort to raise people's incomes to boost their calorie consumption and protein intake should be refocused to include improvements in people's micronutrient intake and dietary

Smallholder poultry farming, Bangladesh.

diversity. With this shift comes the recognition that the pathways through which agricultural development affects hunger and food security are more complex than previously understood.

Taking aim at this challenge is an innovative program in Bangladesh that promotes home gardening, small livestock production, and nutrition education. Helen Keller International (HKI), a nongovernmental organization, worked in partnership with more than 70 local organizations and the Government of Bangladesh to encourage food-insecure households to grow their own micronutrient-rich foods for both home consumption and the market. These homestead food production programs have reached an estimated 5 million individuals and contributed to combating micronutrient deficiencies that can be major causes of diseases such as night blindness, particularly among women and children.

Nutrition education class, Bangladesh

Why did it work?

The successes in agricultural development highlighted here provide insights that are important to those directly and indirectly involved in policy, programs, and investments in agriculture.

Science and technology. Sustained investment in agricultural research and development is vital to developing-country agriculture. The application of science and technology to agricultural development—whether by developing advanced techniques for crop breeding or updating farmers' traditional soil and water management practices—is a common determinant of success. The critical role of long-term public investment in science and technology plays out across the entire developing world and across a range of successes from major food crops to lesser-known crops and to livestock and fisheries.

Complementary investments. Still, science and technology are not enough: sustained public investment in the hardware and software of agricultural development is also critical. This includes public investment in irrigation schemes, rural road networks, rural education, market infrastructure, and regulatory systems. Long-term public investment in the building blocks of agricultural development is a necessary condition for success and is evident in each and every success case.

Private incentives. But even with sustained public investment in science, technology, and complementary investment areas, little can be achieved without the right incentives. By putting policies in place that encourage farmers, entrepreneurs, and companies to invest in agriculture, and by ensuring that markets provide accurate and timely price signals to these private sector actors, the likelihood of success in agricultural development increases.

Cooperation and collaboration. Partnerships among diverse actors in the agricultural sector—research institutes, community-based organizations, private companies, government agencies, and international bodies—are evident in almost all successes. But collaborative interventions are a tricky business and require know-how in managing public and private resources

Millet seed ready for planting, India.

effectively, orchestrating foreign assistance and community resources, and managing relationships among sometimes disparate interest groups.

Timing and planning. Many successes result from good timing, whether by chance or design. In some cases, the time was simply right for the intervention—the technological, economic, social, and political elements were all in place. In other cases, the intervention was adjusted to ensure that the timing was right: gradual reforms were undertaken step-by-step, calculated measurements of the potential gains and losses were carried out, and a strong degree of support was provided to those affected by the reforms.

Experimentation and evolution. Often, successes emerge from localized experiments that allow participants to learn from their mistakes, adapt to changes in the landscape, evolve as the playing field becomes more complex, and pursue incremental, step-by-step approaches to scaling up. Creating space for local experimentation and innovation is a critical means of generating big bangs from incremental changes.

Community involvement. Relatedly, by vesting communities with a stake in ownership of a development process, grassroots participation contributes much to the long-term sustainability of a success. Involving communities and smaller groups in local consultations, policy deliberations, scientific research, and experimentation is all part of building from the bottom-up to achieve success. Similarly, involving local practices, customs, and knowledge in an intervention are the seeds of big successes.

Leadership and dedication. Often, the solutions needed to address agricultural development challenges require dedicated individuals to make the difference—champions to push the issue to the forefront of the public's consciousness, demonstrate what can be done in the face of seemingly insurmountable challenges, or mobilize the political and financial capital to overcome inertia. These champions, both renowned and anonymous, are essential ingredients of success. Creating an environment that encourages leadership on such issues and rewards individuals based on their merit is important to creating success.

What can we learn?

How can the successes of the past help inform and influence agricultural investments that will contribute to substantially reducing hunger in the future? A few reflections are offered here.

Success is not a substitute for strategy. Individual successes at any size or scale must stimulate broader and more sustained processes of national and global success building. But these processes are feasible only if countries pursue good strategies, create supporting policies, and encourage the appropriate levels of investment and experimentation needed to accumulate successes that eventually add up to a sustained success. Without these necessary conditions, successes will likely be scattered, occasional events—outcomes of an unexpected scientific breakthrough or a one-off policy correction.

Success is a process. As such, successes are generated and sustained through experiential processes. This means discovering by doing, learning from mistakes, and adapting to change. The importance of designing an intervention that allows for learning and adaptation can increase the likelihood of success.

Success is recognizable. Sometimes successes emerge only in retrospect, once a substantial amount of time has passed to allow for reflection. But for successes in agricultural development to be recognized as such, they need to be sufficiently supported by strong evidence. Successes in agricultural development—and failures too—need to be systematically documented, examined, and shared so that others can learn lessons, adapt them to different circumstances and contexts, or avoid similar pitfalls.

Success is not unambiguous. In many cases, it is immediately obvious that there is no such thing as an "unequivocal success." Many successes are often accompanied by some type of trade-off.

Unfortunately, this ambiguity may be one reason why agricultural development became such an unpopular topic among both governments and donors in the 1980s, along with other reasons such as donor fatigue, bureaucratic intransigence, and weak private sector responses. As a result, public investment and donor assistance declined precipitously during this period: agricultural research spending stagnated while rural infrastructure development came to a halt in many developing countries.

Yet investments in agricultural development have generated sizable dividends for society, demonstrating that agriculture is not only an important means of reducing poverty, but also a worthwhile investment portfolio.

Homestead food production, Bangladesh.

Looking ahead

Looking to the future, the changing realities of the global food and agriculture system, and the persistence of hunger in the developing world, indicate that more and more frequent successes are needed. Agriculture is increasingly driven by market demand forces, consumer preferences, regulatory scrutiny, and ethical considerations. Agriculture is far more commercial and far more globalized through domestic market growth, international trade, and global finance than ever before. Emerging information, communications, and biological technologies are providing new opportunities for farmers and consumers, while climate change is imposing new constraints on agricultural practices, rural livelihoods, and the resilience of agroecological systems. New demographic concerns are emerging with the continuing HIV/AIDS pandemic, changing age structures in some developing countries, rapid urbanization and rural flight, and growing regional and global migration.

The tools needed to address these evolving realities have changed during the last five decades, but the essentials remain unchanged—increasing the production of, access to, and quality of food to end hunger and feed millions. All of the lessons learned here must be applied and adapted for the future, but with a greater sense of urgency and commitment.

FIGHTING A "SHIFTY ENEMY" The international collaboration to contain wheat rusts

H. J. Dubin and John P. Brennan

Key period: 1955–present
Geographic region: Global

The intervention: The global effort to fight wheat rusts—fungi that can rapidly decimate wheat as it matures in the field—involved a critical breakthrough that brought modern science to bear on the challenges of agricultural development. Nobel Prize Laureate Norman Borlaug, with the eventual backing of policymakers, scientists, and philanthropists, catalyzed this effort by breeding rust-resistant wheat varieties in Mexico using innovative research methods. As a result, about 117 million hectares of land under wheat cultivation were protected from the fungi, directly ensuring the food security of 60 to 120 million rural households and many more millions of consumers.

BREEDING AN "AMAIZING" CROP Improved maize in Kenya, Malawi, Zambia, and Zimbabwe

Melinda Smale and T. S. Jayne

Key period: 1965–1990
Geographic region: Kenya, Malawi, Zambia, Zimbabwe

The intervention: Sustained investments in innovative breeding programs, dedicated scientists, and supportive public policies drove the development and spread of more productive maize that translated into better livelihoods for millions of farm households. By expanding access to modern (improved) maize seeds among smallholder farmers, yields multiplied several-fold and contributed significantly to improving food production and food security in the region. While the fiscal burdens of state-led marketing and credit policies rendered the growth unsustainable, by 2000–05, maize, most of it modern maize, covered more than three-quarters of the land under cereal cultivation in the four countries.

TRANSFORMING AGRICULTURE The Green Revolution in Asia

Peter B. R. Hazell

Key period: 1965–1985
Geographic region: Bangladesh, China, India, Indonesia, Republic of Korea, Malaysia, Myanmar, Nepal, Pakistan, Philippines, Sri Lanka, Thailand, Vietnam

The intervention: The breeding of improved rice and wheat varieties, combined with the expanded use of fertilizers and other chemical inputs, irrigation, and supportive public policies led to dramatic yield and production increases in Asia beginning in the late 1960s. As the Green Revolution spread rapidly across Asia, the resultant increases in food production—including cereal yield increases of 100 percent between 1970 and 1995—pulled the region back from the edge of famine. Within 20 years, cereal production had doubled and per capita income increased 190 percent, improving livelihoods for an estimated 1.8 billion people in rural Asia and saving large areas of fragile lands from conversion to cropping.

RESISTING VIRUSES AND BUGS Cassava in Sub-Saharan Africa

Felix Nweke

Key period: 1971–1989
Geographic region: Sub-Saharan Africa

The intervention: Two major control programs were designed to combat serious threats to cassava production in Sub-Saharan Africa—the cassava mosaic disease and the cassava mealybug. These programs played a critical role in raising cassava yields beginning in the 1970s, turning cassava into a cash crop that is now spreading throughout Africa. In the early 1970s, the introduction of bio-control strategies to destroy mealybug infestations reduced yields losses by 2.5 tons per hectare. In the late 1970s, the introduction of improved, disease-resistant varieties controlled cassava mosaic disease while contributing to yield increases of 40 percent. These two programs played a particularly critical role in countries such as Nigeria and Ghana, and have contributed to improvements in food security for at least 29 million people.

SEEING THE FOREST THROUGH THE TREES
Community forestry in Nepal

Hemant Ojha, Lauren Persha, and Ashwini Chhatre

Key period: 1978–present
Geographic region: Nepal

The intervention: Policy reforms have facilitated innovation in participatory, community-based governance of Nepal's forests, an important natural resource, over the last three decades. The country moved from a conservation-focused agenda to a more broad-based strategy encompassing forest use, enterprise development, and livelihoods improvement. By combining forestry and agriculture within a framework that includes experiential learning, a strong civil society network, and progressive policymaking, Nepal has developed community forestry into an important contributor to food security for an estimated 1.7 million rural households. As of 2009, one-third of Nepal's population was directly managing over one-fourth of Nepal's forest area, using it as a source of raw materials, cash income, and employment.

INNOVATING IN THE PAMPAS
Zero-tillage soybean cultivation in Argentina

Eduardo Trigo, Eugenio Cap, Valeria Malach, and Federico Villarreal

Key period: 1989–present
Geographic region: Argentina

The intervention: The introduction of zero-tillage cultivation techniques among Argentine farmers has contributed to a significant increase in global supplies of soybean, an essential food and feed crop, elevating Argentina to a leading position in global soybean production. An innovative partnership among farmers, researchers, extension workers, and private companies was formed in the late 1980s to promote zero tillage. Nearly 22 million hectares of land were brought under this resource-conserving cultivation practice between 1991 and 2008. The use of zero tillage, along with the introduction of herbicide-resistant soybean varieties and other factors, improved soil fertility by reversing decades of erosion, created an estimated 200,000 new agricultural jobs, and provided the international market with new supplies of soybeans that contributed to keeping global food prices low.

RE-GREENING THE SAHEL
Farmer-led innovation in Burkina Faso and Niger

Chris Reij, Gray Tappan, and Melinda Smale

Key period: 1980–present
Geographic region: Burkina Faso and Niger

The intervention: The rediscovery and diffusion of traditional agroforestry, water, and soil management practices in Burkina Faso and Niger has transformed large swathes of the region's arid landscape into productive agricultural land. In Burkina Faso's Central Plateau, the rehabilitation of between 200,000 and 300,000 hectares translated into roughly 80,000 tons of additional food per year, enough to sustain about half a million people in the region. In southern Niger, farmer investments in agriculture are estimated to have transformed approximately 5 million hectares of land, improving food security for at least 2.5 million people.

LEAVING THE PLOW BEHIND
Zero-tillage rice–wheat cultivation in the Indo-Gangetic Plains

Olaf Erenstein

Key period: 1995–present
Geographic region: India, Pakistan

The intervention: An estimated 620,000 wheat farmers in northern India have benefited significantly from the introduction of crop management techniques known as zero-tillage cultivation. In this practice, the seeds are planted in unplowed fields in order to conserve soil fertility, economize on scarce water, reduce land degradation, and lower production costs. Varying forms of the technique have been adopted over an estimated 1.76 million hectares of wheat, particularly in the Indian states of Haryana and Punjab, with average income gains amounting to US$180–340 per household.

PUMPING UP PRODUCTION
Shallow tubewells and rice in Bangladesh

Mahabub Hossain

Key period: 1989–present
Geographic region: Bangladesh

The intervention: Market liberalization in Bangladesh during the mid-1980s eased restrictions on the importation and sale of irrigation equipment, primarily low-lift power pumps and shallow tube wells. These reforms stimulated the growth of irrigated rice farming during the dry season, accounting for 90 percent of the increase in rice production in Bangladesh between 1988 and 2007 and benefiting about 22 million people per year. With this growth in rice production came a decline in real rice prices, benefiting food-insecure households in particular and ultimately significantly reducing poverty.

IMPROVING CROPS FOR ARID LANDS
Pearl millet and sorghum in India

Carl E. Pray and Latha Nagarajan

Key period: mid-1960s–present
Geographic region: India

The intervention: Diffusion of improved pearl millet and sorghum was achieved through long-term investments by the Indian government, state governments, and the international agricultural research system. This effort contributed vitally to reducing food insecurity in India's arid and semi-arid tropics, where a majority of the country's poor are still concentrated. National average yields of sorghum and pearl millet have increased by up to 85 percent over the last four decades, and almost 80 percent of sorghum and pearl millet areas are now sown with high yielding varieties. The emergence of private seed companies further expanded access to these improved varieties among the 6 to 9 million smallholder households cultivating these crops.

PUSHING THE YIELD FRONTIER
Hybrid rice in China

Jiming Li, Yeyun Xin, and Longping Yuan

Key period: 1977–present
Geographic region: China

The intervention: Through the efforts of plant scientists, seed producers, extension agents, and farmers, China became the first country to develop and commercialize hybrid rice. With average yields that exceed those of other cultivated rice varieties by between 15 and 31 percent, hybrid rice has allowed China to feed an additional 60 million people per year, while reducing the land allocated to rice production by 14 percent since 1978. Hybrid rice now accounts for 63 percent of all land under rice cultivation and has spawned a vibrant rice seed industry.

NAVIGATING THROUGH REFORMS
Cotton reforms in Burkina Faso

Jonathan Kaminski, Derek Headey, and Tanguy Bernard

Key period: 1992–2006
Geographic region: Burkina Faso

The intervention: In an effort to revamp an inefficient state-led cotton development strategy, Burkina Faso pursued a gradual and sequenced reform process that, first, strengthened the role of cotton farmers' groups and, second, partially liberalized input and output markets in the cotton sector. Partly as a result of these reforms, and despite low cotton prices worldwide, Burkina Faso emerged as the leading African exporter of cotton in 2006, with a threefold increase in production since the early 1990s. Reforms and growth in the cotton sector affected over 175,000 households (about 1.8 million people) and generated an estimated 235,000 new full-time jobs in the agricultural sector.

UNLOCKING THE MARKET
Fertilizer and maize in Kenya

Joshua Ariga and T. S. Jayne

Key period: 1990–2007
Geographic region: Kenya

The intervention: In Kenya, policy reforms in the early-1990s contributed to the rapid growth of private investment in fertilizer and maize marketing, resulting in a dramatic reduction in farmers' time, effort, and costs associated with purchasing fertilizer and selling surplus maize production. The average distance to the nearest fertilizer retailer for small farmers has decreased by more than half between 1997 and 2007, reflecting a large expansion in the number of fertilizer retailers serving smallholder areas. The distance traveled by farmers to sell their maize to private traders also declined over this period. The proportion of small-scale farmers using fertilizer on maize during the main growing season rose from 56 percent in 1996 to 70 percent in 2007, contributing to improved productivity and availability of maize, a vitally important staple crop for Kenyan consumers.

CONQUERING THE CATTLE PLAGUE
The global effort to eradicate rinderpest

Peter Roeder and Karl Rich

Key period: 1950–2001
Geographic region: Global

The intervention: Concerted effort by national veterinary services in both industrialized and developing countries, aided by international organizations, has brought the once dreaded rinderpest livestock virus to the point of extinction. Control programs in the last 20 years have contributed significantly to the eradication of rinderpest, using vaccine innovations and new epidemiological and surveillance tools based on participatory techniques. These programs have protected some 40 million poor livestock keepers from experiencing major losses in milk, meat, and hide production, as well as losses of household incomes and assets. The fact that the virus is no longer circulating in domesticated or wild animals anywhere in the world is a remarkable achievement, on a par with the eradication of smallpox in the human population—the only other case of global eradication of an infectious disease.

COUNTING ON BEANS
Mungbean improvement in Asia

Subramanyam Shanmugasundaram, J.D.H. Keatinge, and Jacqueline d'Arros Hughes

Key period: 1980s–present
Geographic region: Bangladesh, Bhutan, China, India, Nepal, Myanmar, Pakistan, Sri Lanka, and Thailand

The intervention: Partly as a result of a cross-country research program that involved farmers in the innovation process, the cultivation of mungbean—a little-known pulse crop that is high in protein, iron, and other micronutrients, and useful in maintaining soil fertility—has grown dramatically since the 1980s. The release of mungbean varieties with traits such as higher yields, short maturity times, and other characteristics tailored to different local environments contributed to yield gains of 28 to 55 percent among an estimated 1.5 million farmers. This was a key factor in the 35 percent increase in Asian mungbean production between 1985 and 2000 that accompanied a significant increase in global consumption during the same period.

CONNECTING THE MILK GRID
Smallholder dairy in India

Kenda Cunningham

Key period: 1970–1996
Geographic region: India

The intervention: Operation Flood, a dairy development program that ran from 1970 to 1996 in India, laid the foundation for the integration and growth of the national dairy industry. In the process of linking India's major cities with rural cooperatives, Operation Flood brought significant technological advances into the rural milk sector, commercialized smallholder dairy production, and transformed the policy environment in support of dairy industry growth. Of the program's 9 million direct beneficiaries, 73 percent were small, marginal, and landless farmers who saw their incomes double from this intervention. Since 1970, India's dairy industry has steadily grown. India has been transformed from a dairy-importing nation to a top producer of milk in the world; millions of consumers have benefited from improved access to milk and other dairy products.

FARMING THE AQUATIC CHICKEN
Improved tilapia in the Philippines

Sivan Yosef

Key period: 1988–1997
Geographic region: Philippines

The intervention: The Genetic Improvement of Farmed Tilapia (GIFT) project served as a launching point for tropical finfish genetic improvement around the world. Based on selective breeding of Nile tilapia, the GIFT project succeeded in producing tilapia that grows faster and has a higher survival rate, thus increasing fish yields dramatically. Between 1990 and 2007, tilapia production in the Philippines increased by 186 percent, while costs of producing tilapia declined by 32 to 35 percent. The stable, low price of GIFT-derived fish has made it an extremely popular protein source among an estimated 19 to 23 million poor consumers. These fish strains now comprise 68 percent of total tilapia seed produced in the Philippines.

EXITING FROM COLLECTIVE AGRICULTURE
Land-tenure reform in Vietnam

Michael Kirk and Nguyen Do Anh Tuan

Key period: 1988-1993
Geographic region: Vietnam

The intervention: Vietnamese land tenure policy reforms that were part of its wider economic reforms under the Doi Moi program enabled the country's transition towards a market economy. The decollectivization of agricultural production and improvement of land tenure security—along with the liberalizing of markets and promotion of new economic incentives—played a critical role in accelerating agricultural growth, increasing food security, and reducing poverty throughout the country. Between 1986 and 2005, Vietnam's agricultural growth rate averaged 3.8 percent per year, and Vietnam became a major global exporter of rice, coffee, and other crops during the period.

CROSSING THE RIVER WHILE FEELING THE ROCKS
Land-tenure reform in China

John W. Bruce and Zongmin Li

Key period: 1978–1984
Geographic region: China

The intervention: Beginning in the late 1970s, China undertook a series of policy reforms that transformed the country's agricultural sector and reduced hunger on a scale unparalleled in history. The reforms effectively reintroduced household-based farming after more than 30 years of collective agriculture, returning more than 95 percent of China's farmland to some 160 million farm households and giving farmers the incentive to sell their surplus farm production to market. The reforms contributed to increasing rural incomes by 137 percent, reducing rural poverty by 22 percent, and increasing grain production by 34 percent. Gains in on-farm efficiency also led to a substantial increase in the rural labor force available for non-agricultural employment—a shift that fueled China's rapid industrialization in recent decades.

DIVERSIFYING INTO HEALTHY DIETS
Homestead food production in Bangladesh

Lora Iannotti, Kenda Cunningham, and Marie Ruel

Key period: 1990–present
Geographic region: Bangladesh

The intervention: Interventions that address poor diet quality (and related deficiencies of vitamin A, zinc, iron, and other nutrients) are important for achieving full food security in vulnerable populations. The homestead food production (HFP) program, introduced in Bangladesh by Helen Keller International, promotes an integrated package of home gardening, small livestock production, and nutrition education. This program has reached an estimated 5 million individuals, contributing to improving the micronutrient intakes of women and children, empowering women, and supporting community development.

Advisory Committee

Harris Mule, Committee Co-Chair, Former Permanent Secretary, Ministry of Finance, Kenya

Raul Montemayor, Committee Co-Chair, General Secretary, Federation of Free Farmers Cooperatives, Inc., Philippines

Chris Dowswell, Executive Director, Sasakawa Africa Association, Mexico

Mahabub Hossain, Executive Director, Bangladesh Rural Advancement Committee, Bangladesh

Isatou Jallow, Chief, Women, Children and Gender Policy, World Food Programme, Italy

Marina Joubert, Science Communication Editor, Southern Science, South Africa

Ruth Levine, Vice President, Programs and Operations, Center for Global Development, USA

Xiaopeng Luo, Professor, China Academy for Rural Development, Zhejiang University, China

Stephen Muliokela, Executive Director, Golden Valley Agricultural Research Trust, Zambia

Raj Paroda, Executive Secretary, Asia-Pacific Association of Agricultural Research Institutions, Thailand

Christie Peacock, Chief Executive, Farm Africa, United Kingdom

Prabhu Pingali, Deputy Director, Agricultural Development, The Bill & Melinda Gates Foundation, USA

Martín Piñeiro, Director, GrupoCeo, Argentina

Papa Seck, Director General, Africa Rice Center, Benin

Camilla Toulmin, Director, International Institute for Environment and Development, United Kingdom

Ajay Vashee, President, International Federation of Agricultural Producers, Zambia

Joachim von Braun, Ex-Officio Member, Director General, International Food Policy Research Institute, USA

IFPRI Project Team

Rajul Pandya-Lorch, Project Leader—Chief of Staff, Director General's Office, and Head, 2020 Vision Initiative

David J. Spielman, Project Research Lead—Research Fellow, Knowledge, Capacity, and Innovation Division

Klaus von Grebmer, Project Communications Lead—Director, Communications Division

Kenda Cunningham, Research Support—Senior Research Assistant, Director General's Office

Sivan Yosef, Research Support—Senior Research Assistant, Director General's Office

International Food Policy Research Institute (IFPRI)

The International Food Policy Research Institute, established in 1975, seeks sustainable solutions for ending hunger and poverty. IFPRI is one of 15 centers supported by the Consultative Group on International Agricultural Research (CGIAR), an alliance of 64 governments, private foundations, and international and regional organizations. The Institute's research and capacity-strengthening and communications activities are made possible by its financial contributors and partners. IFPRI gratefully acknowledges the generous unrestricted funding from Australia, Canada, China, Finland, France, Germany, India, Ireland, Italy, Japan, Netherlands, Norway, South Africa, Sweden, Switzerland, United Kingdom, United States, and World Bank.

2020 Vision Initiative

"2020 Vision for Food, Agriculture, and the Environment" is an IFPRI initiative to develop a shared vision and consensus for action on how to meet future world food needs while reducing poverty and protecting the environment. Through the 2020 Vision Initiative, IFPRI is bringing together divergent schools of thought on these issues, generating research, and identifying recommendations.

Millions Fed Project

"Millions Fed: Proven Successes in Agricultural Development" is a project led by IFPRI, with support from The Bill & Melinda Gates Foundation, to identify interventions in agricultural development that have substantially reduced hunger and poverty; to document evidence about where, when, and why these interventions succeeded; to learn about the key drivers and factors underlying success; and to share lessons to help inform better policy and investment decisions in the future.

This booklet offers highlights and lessons from the project and the case studies. More detailed narratives appear in a book called *Millions Fed: Proven Successes in Agricultural Development* and complete technical background papers and other information appear on the Millions Fed website (www.ifpri.org/millionsfed).

Authors

David J. Spielman (d.spielman@cgiar.org) is a research fellow with the Knowledge, Capacity, and Innovation Division of the International Food Policy Research Institute and is based in Addis Ababa, Ethiopia.

Rajul Pandya-Lorch (r.pandya-lorch@cgiar.org) is Chief of Staff in the Director General's Office and Head of the 2020 Vision Initiative of the International Food Policy Research Institute.

Made in the USA
Monee, IL
05 November 2026

67379778R00019